HOW FAST?

by Nicholas Harris

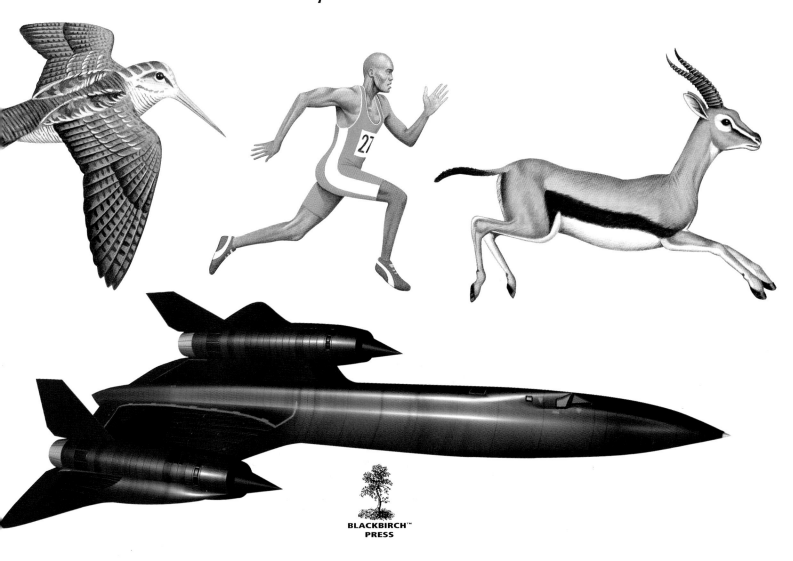

BLACKBIRCH PRESS

THOMSON
GALE

Detroit • New York • San Diego • San Francisco • Cleveland • New Haven, Conn. • Waterville, Maine • London • Munich

6   3 inches per hour – 5.5 miles per hour
Amoeba • Garden snail • Sloth • Giant tortoise • Cockroach

8   3 – 9 miles per hour
Person walking • Sprinting swimmer • American woodcock • Polar bear swimming • Rat • Benz 1885 car • Crocodile running

10   11 – 15 miles per hour
Pig • Honeybee • *Santa Maria* • Marathon runner • *Tyrannosaurus rex* • Sea trout

12   18 – 25 miles per hour
Dragonfly • Female sprinter • Stagecoach • Male sprinter • African elephant • *Titanic*

14   28 – 35 miles per hour
Speed ice skater • *Rocket* • Wright *Flyer* • Rhinoceros • Tiger shark • Polar bear running • Racing yacht

16   40 – 52 miles per hour
Red kangaroo • Greyhound • Racehorse • Racing cyclist • Brown hare • Ostrich • Bluefin tuna • Thomson's gazelle • Submarine • Windsurfer

ABBREVIATIONS
in   inches
ft   feet
cm   centimeter
m   meter
km/h   kilometers per hour

Where the metric equivalent is given, the conversion is approximate.

CONTENTS

18 55 – 85 miles per hour
Pronghorn antelope • Skateboarder • Racing pigeon • Tornado • Mallard duck • Cheetah • *Hindenburg* airship • Luge

20 105 – 163 miles per hour
Table tennis smash • Spine-tailed swift • Peregrine falcon • *Mallard* steam locomotive • Tennis serve • Downhill skier • Powerboat • Light aircraft

22 180 – 342 miles per hour
Golf ball drive • Jai alai ball throw • Racing car • Lynx helicopter • *Spirit of Australia* • TGV • *Easyrider* motorcycle • Longbow arrow

24 580 – 1,523 miles per hour
Revolver bullet • Boeing 747 • Bell X-1 • *ThrustSSC* • Concorde • F-16

26 2,046 – 4,518 miles per hour
Rifle bullet • Moon • Lockheed SR-71 • Surface-to-air missile • Anti-tank projectile • V2 rocket • X-15

28 17,640 – 107,106 miles per hour
Space Shuttle • Apollo 10 • Voyager 2 • Meteoroids • Earth • Mercury

30 Index

AN AMOEBA is neither a plant nor an animal. It is a protist. A protist is a microscopic living thing made up of only one cell (the building block of life). An amoeba is like a plastic bag full of jello. It moves about by flowing like a liquid. It can travel about three inches (8 cm) in an hour—not a bad speed for a creature itself only a fraction of an inch long.

A snail moves by sliding over slime that seeps out of its single foot. When it needs to, it can quickly protect itself from danger by disappearing into its shell.

Amoeba
3 in (8 cm) per hour

Garden snail
165 ft (50 m) per hour

0.0005 mph

0.03 mph

3 inches/hr – 5.5 mph

Three-toed sloth
620 ft (190 m) per hour

Galápagos giant tortoise
1200 ft (365 m) per hour

Cockroach
2.2 mph (3.5 km/h)

0.1 mph   1.5 mph   2 mph

Sloths spend so much of their lives in the trees, they cannot walk normally on the ground. Instead, they hang upside down from the branches. They use their curved, hooklike claws to grip the branches. They eat, mate, and give birth hanging upside down. Even their fur grows downwards. Although they move extremely slowly, sloths can strike out quickly with their claws when they are threatened.

Like snails, tortoises have a shell to protect them from attack. So they do not have to move very quickly at all. The Galápagos giant tortoise may live for more than 150 years.

Compared to snails and tortoises, cockroaches are sprinters. They run away on their long legs whenever danger threatens.

A SPRINTING SWIMMER can race through the water faster than you can walk. But he or she would soon be overtaken by a polar bear. Speedy over land or ice (it can easily outrun a reindeer), the polar bear is also an excellent swimmer. It needs to be, so it can cross from one floating slab of ice to the next. A polar bear can travel as fast through water as a rat can move on land. A bear uses its powerful front limbs to pull its large body along through the icy water.

The American woodcock is believed to be the slowest-flying of all birds. It travels just quickly enough to keep it from falling out of the sky! It is a shy bird that feeds mainly on earthworms.

American woodcock
5 mph (8 km/h)

Sprinting swimmer
5 mph (8 km/h)

Walking pace
3 mph (5 km/h)

3 mph   4 mph   5 mph

3 – 9 mph

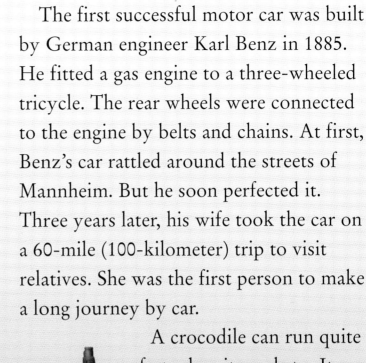

Rat
mph (10 km/h)

Benz 1885 car
(early petrol
engine car)
8 mph (13 km/h)

The first successful motor car was built by German engineer Karl Benz in 1885. He fitted a gas engine to a three-wheeled tricycle. The rear wheels were connected to the engine by belts and chains. At first, Benz's car rattled around the streets of Mannheim. But he soon perfected it. Three years later, his wife took the car on a 60-mile (100-kilometer) trip to visit relatives. She was the first person to make a long journey by car.

A crocodile can run quite fast when it needs to. It will push itself along on its belly, whipping its tail from side to side. Some smaller kinds of crocodile can even move as fast as a rabbit!

	THE FASTEST REPTILES	
1	Pacific leatherback turtle	22 mph
2	Spiny-tailed iguana	21.7 mph
3	Zebra-tailed lizard	18 mph
4	Six-lined racerunner	17 mph
5	Great Basin whiptail	15 mph
6	Black mamba snake	14 mph

lar bear
imming
(10 km/h)

Crocodile running
8.7 mph (14 km/h)

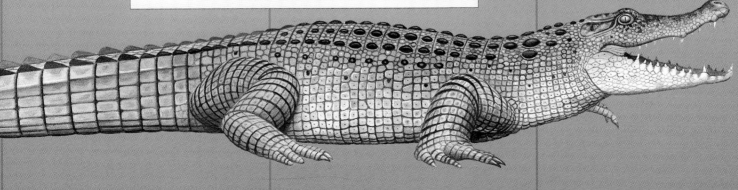

7 mph   8 mph

FOR HIS voyage across the Atlantic Ocean in 1492, Christopher Columbus took three ships. They were the *Niña*, the *Pinta,* and his flagship, the *Santa Maria*. All three were three-masted vessels called caravels. The *Santa Maria* was not very fast. It sailed only a little faster than a pig could run. A long-distance runner could have gone more quickly. The *Santa Maria* successfully completed the journey from Spain to America. It eventually ran aground on a coral reef on Christmas Eve, 1492.

Honeybee 11 mph (18 km/h)

Pig 11 mph (18 km/h)

9 mph 10 mph 11 mph

11 – 15 mph

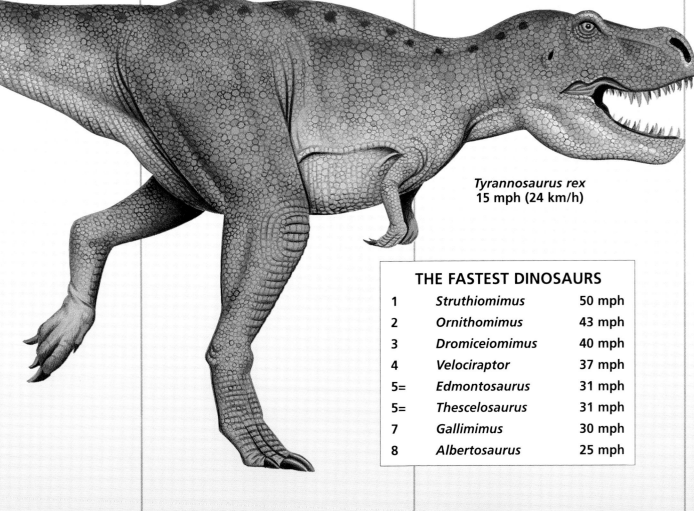

Tyrannosaurus rex
15 mph (24 km/h)

	THE FASTEST DINOSAURS	
1	*Struthiomimus*	50 mph
2	*Ornithomimus*	43 mph
3	*Dromiceiomimus*	40 mph
4	*Velociraptor*	37 mph
5=	*Edmontosaurus*	31 mph
5=	*Thescelosaurus*	31 mph
7	*Gallimimus*	30 mph
8	*Albertosaurus*	25 mph

Sea trout
15 mph (24 km/h)

Santa Maria
Christopher Columbus's flagship
12 mph (20 km/h)

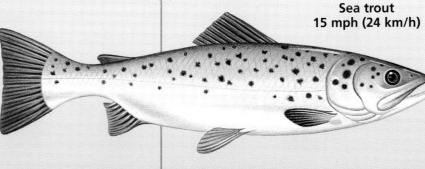

Marathon runner
13 mph (21 km/h)

Tyrannosaurus rex was not the quickest of dinosaurs. But it may have been able to sprint quite fast for short bursts. It may have chased down its prey—perhaps a plant-eating dinosaur that was too sick, too old, or too young to keep up with the rest. The great predator would then finish off its victim with a killer bite.

13 mph   14 mph   15 mph

Dragonfly
18 mph (29 km/h)

RUNNING AT world record pace, a top male athlete could cover 100 meters (about 300 ft) in just under 10 seconds. The world's fastest women would not be far behind. Both male and female sprinters could easily beat the fastest flying insects, such as dragonflies. These creatures use their speed to dart across water and pick off other insects. They can make rapid changes of direction— they can even fly backwards!

16 mph

18 mph

18 – 25 mph

Female sprinter
20.5 mph
(33 km/h)

Male sprinter
22 mph
(36 km/h)

Over a short distance, sprinters could even keep up with a stagecoach traveling at full speed. They would be overtaken, however, by a stampeding African elephant. The famous ocean liner *Titanic*, which sank on its first voyage in 1912, steamed across the ocean at about the same speed as a fast elephant.

African elephant
25 mph (40 km/h)

Titanic ocean liner
25 mph (40 km/h)

22 mph

24 mph

**Wright *Flyer 1*
first airplane
30 mph (48 km/h)**

WHEN the locomotive was invented in the early 19th century, it became possible for people to travel faster on land than by horse. The first successful locomotive, the *Rocket*, was built by British engineer George Stephenson in 1829. People soon chose to travel by rail rather than by road carriage.

At the beginning of the next century powered aircraft were invented. The first successful flight was made by American inventors Orville and Wilbur Wright on December 17, 1903.

**Speed ice skater
28 mph (45 km/h)**

***Rocket*
early steam
locomotive
29 mph (47 km/h)**

26 mph   28 mph   30 mph

28 – 35 mph

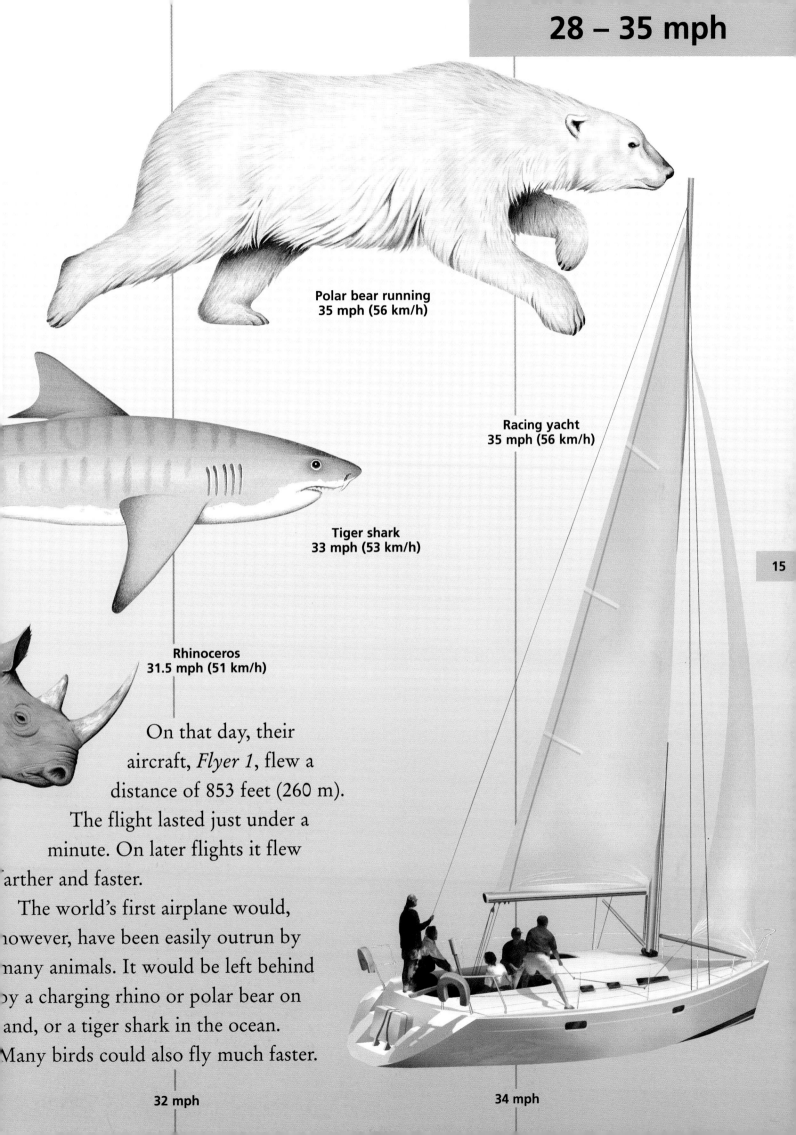

Polar bear running
35 mph (56 km/h)

Racing yacht
35 mph (56 km/h)

Tiger shark
33 mph (53 km/h)

Rhinoceros
31.5 mph (51 km/h)

On that day, their aircraft, *Flyer 1*, flew a distance of 853 feet (260 m). The flight lasted just under a minute. On later flights it flew farther and faster.

The world's first airplane would, however, have been easily outrun by many animals. It would be left behind by a charging rhino or polar bear on land, or a tiger shark in the ocean. Many birds could also fly much faster.

32 mph

34 mph

HERE ARE SOME very fast animals. Each is famous for its speed, either around the racecourse or in the wild. To go as fast them (without traveling in a motorized vehicle), a person would have to cycle at world-record speeds, ride a champion racehorse, or learn to become an extremely skillful windsurfer.

A kangaroo, hare, ostrich, and gazelle are all perfectly built for swift movement across open ground. The kangaroo

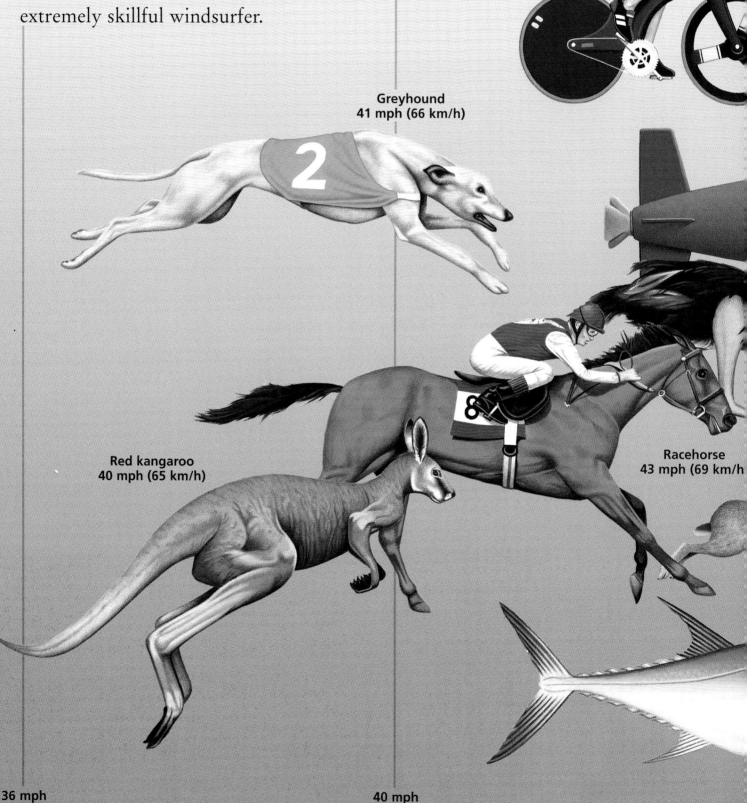

Racing cyclist
44 mph (71 km/h)

Greyhound
41 mph (66 km/h)

Red kangaroo
40 mph (65 km/h)

Racehorse
43 mph (69 km/h)

36 mph

40 mph

40 – 52 mph

ounds forward on two legs. For long distances, this saves more energy than running on four legs.

The ostrich is the fastest animal on two legs. It lives in the grasslands of Africa. It uses its powerful legs and large, two-toed feet to run away from its enemies at a top speed of 45 mph (72 km/h).

In the water, the streamlined tuna fish can almost keep up with one of the fastest types of submarine. It can burst forward over short distances as it chases after prey: fish, crustaceans, and squid.

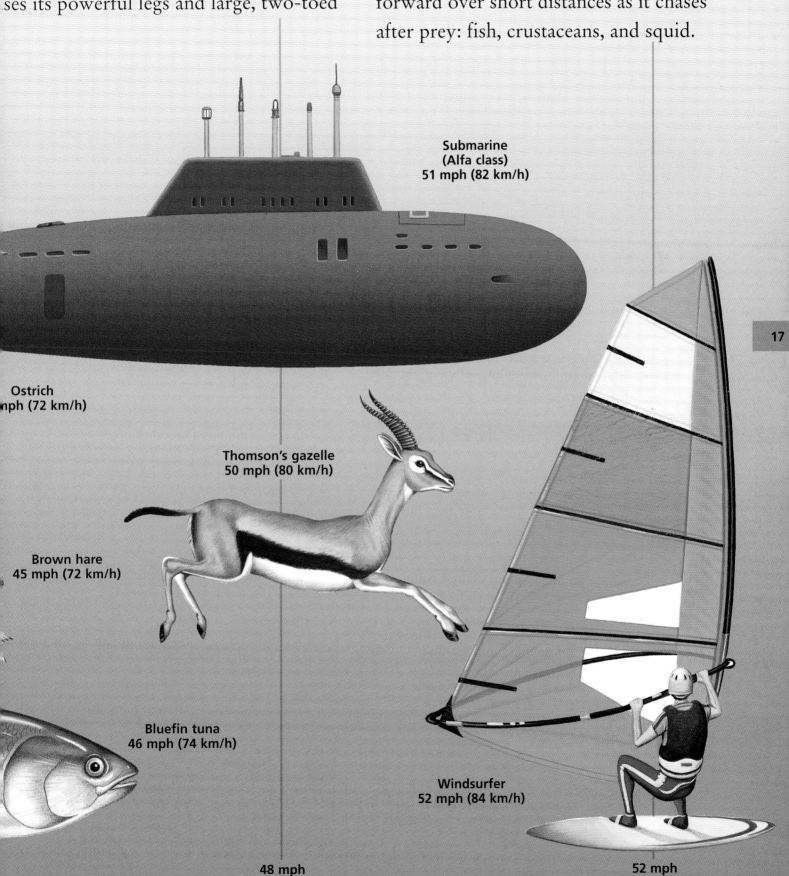

Submarine (Alfa class) 51 mph (82 km/h)

Ostrich
mph (72 km/h)

Thomson's gazelle 50 mph (80 km/h)

Brown hare
45 mph (72 km/h)

Bluefin tuna
46 mph (74 km/h)

Windsurfer
52 mph (84 km/h)

48 mph

52 mph

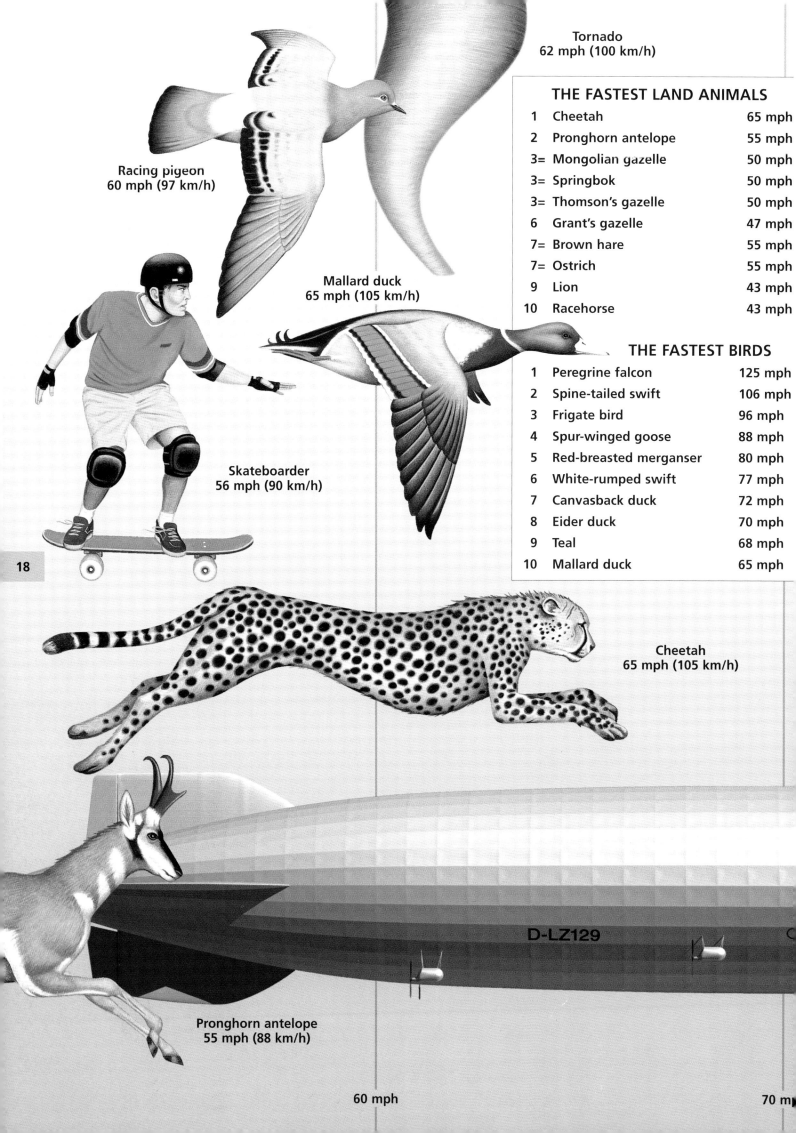

55 – 85 mph

FOR SHORT BURSTS, the cheetah can run faster than any other land animal in the world. With its long legs and lean, muscular body, it is perfectly built for speed. The cheetah uses its speed to catch its prey: antelope, hares, and young ostriches. When a cheetah reaches its victim, it takes it down with a swift bite to the throat.

The pronghorn antelope is probably the fastest land animal over short distances. It needs its speed to escape danger on the wide open grasslands of North America. A four-day-old pronghorn is able to run faster than a human.

Tornadoes are the most powerful storms. The twisting column of air can have wind speeds of more than 250 mph (about 400 km/h). A tornado can itself travel faster than 60 mph (about 100 km/h). It destroys everything in its path.

The great airships of the 1930s were more than 650 feet (about 200 m) long. At their top speed of about 80 mph (130 km/h), they were faster than most birds. But they would be overtaken by an Olympic luge champion, riding a tiny sled flat on his or her back less than three inches (8 cm) above the ice.

Luge
85 mph (137 km/h)

***Hindenburg* airship**
81 mph (130 km/h)

80 mph

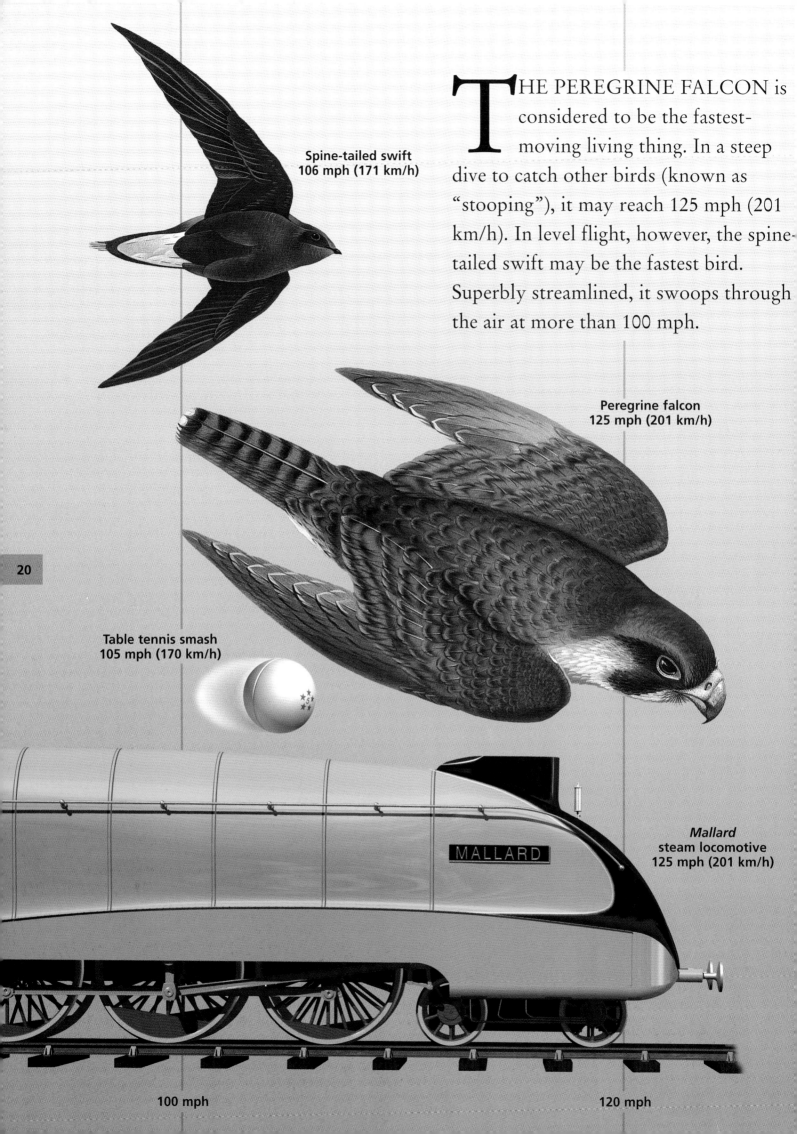

105 – 163 mph

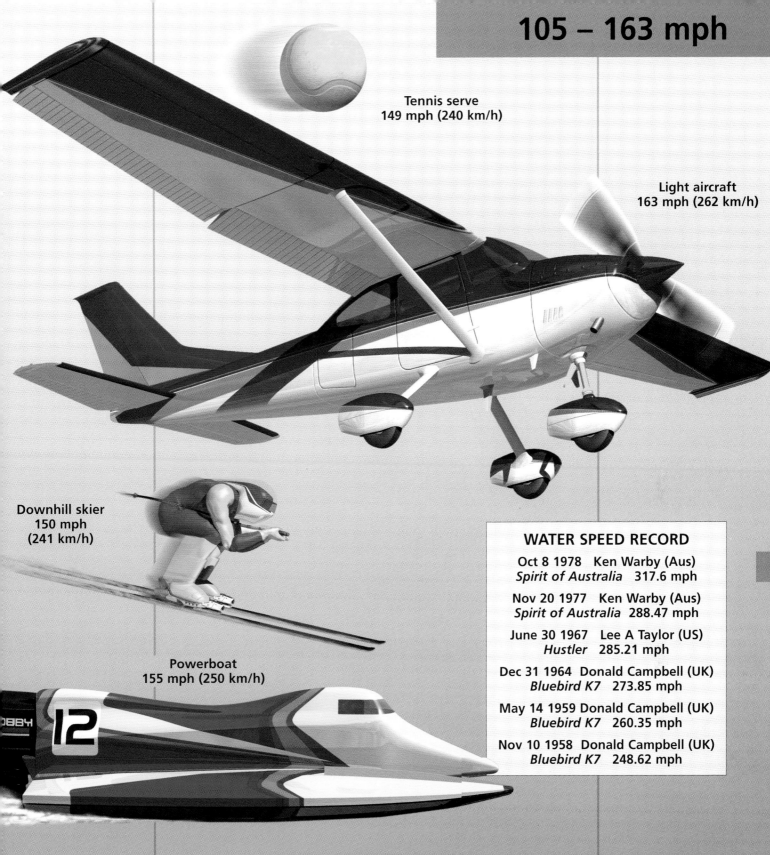

Tennis serve
149 mph (240 km/h)

Light aircraft
163 mph (262 km/h)

Downhill skier
150 mph
(241 km/h)

Powerboat
155 mph (250 km/h)

WATER SPEED RECORD

Oct 8 1978   Ken Warby (Aus)
Spirit of Australia   317.6 mph

Nov 20 1977   Ken Warby (Aus)
Spirit of Australia   288.47 mph

June 30 1967   Lee A Taylor (US)
Hustler   285.21 mph

Dec 31 1964   Donald Campbell (UK)
Bluebird K7   273.85 mph

May 14 1959 Donald Campbell (UK)
Bluebird K7   260.35 mph

Nov 10 1958   Donald Campbell (UK)
Bluebird K7   248.62 mph

In 1938, the steam locomotive *Mallard* set a record over a distance of 400 meters (1312 ft) that stands to this day. The locomotive was fitted with a special streamlined casing.

The water speed record is held by the *Spirit of Australia (see page 23)*. The top speed of most powerboats is about half that. The roughness of the water's surface makes it more difficult to go at high speeds on water. This is much less of a problem moving on ice or through the air.

140 mph

160 mph

Golf ball drive
180 mph (290 km/h)

Jai alai ball, thrown using long, curved basket strapped to the wrist
188 mph (303 km/h)

TO TRAVEL at the speeds shown here, you would have to be behind the wheel of a very fast vehicle. Even a Formula 1 racing car driven at top speed would be beaten by the French TGV *(Train à Grande Vitesse)*. The TGV once clocked a record-breaking 320 mph (515 km/h) on a test run.

Lynx helicopter
248 mph (400 km/h)

Formula 1 racing car
230 mph (370 km/h)

200 mph   250 mph

180 – 342 mph

Helicopters are well known for their ability to take-off and land vertically. Some types can also move through the air extremely fast. A racing car could not keep up with a Lynx flying at full speed. The Lynx, however, would be overtaken by a dragster, the *Spirit of Australia* current holder of the water speed record), and the world's fastest motorcycle.

All these speedy extreme machines would, however, be beaten for speed by an arrow fired from a longbow using just human muscle power!

Spirit of Australia
water speed record holder
317.6 mph (511.11 km/h)

Dragster car
309 mph (497 km/h)

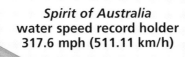

Longbow arrow
342 mph (550 km/h)

Easyrider streamliner
(fastest motorcycle)
322.7 mph (519.6 km/h)

Train à Grande Vitesse
(TGV)
320 mph (515 km/h)

TIME TAKEN FOR CROSSING THE ATLANTIC OCEAN	
Mayflower (1620)	66 days
Dreadnought clipper (1853)	16 days
Mauretania ocean liner (1909)	4 days 10 hrs
Spirit of St. Louis plane (1927)	33 hrs 30 mins
Hindenburg airship (1936)	52 hrs
Boeing 314 airliner (1939)	27 hrs 35 mins
Constellation airliner (1947)	16 hrs 30 mins
Boeing 707 airliner (1968)	6 hrs 40 mins
Concorde (1974)	2 hrs 56 mins
Lockheed SR-71 (1974)	1 hr 55 mins

300 mph   350 mph

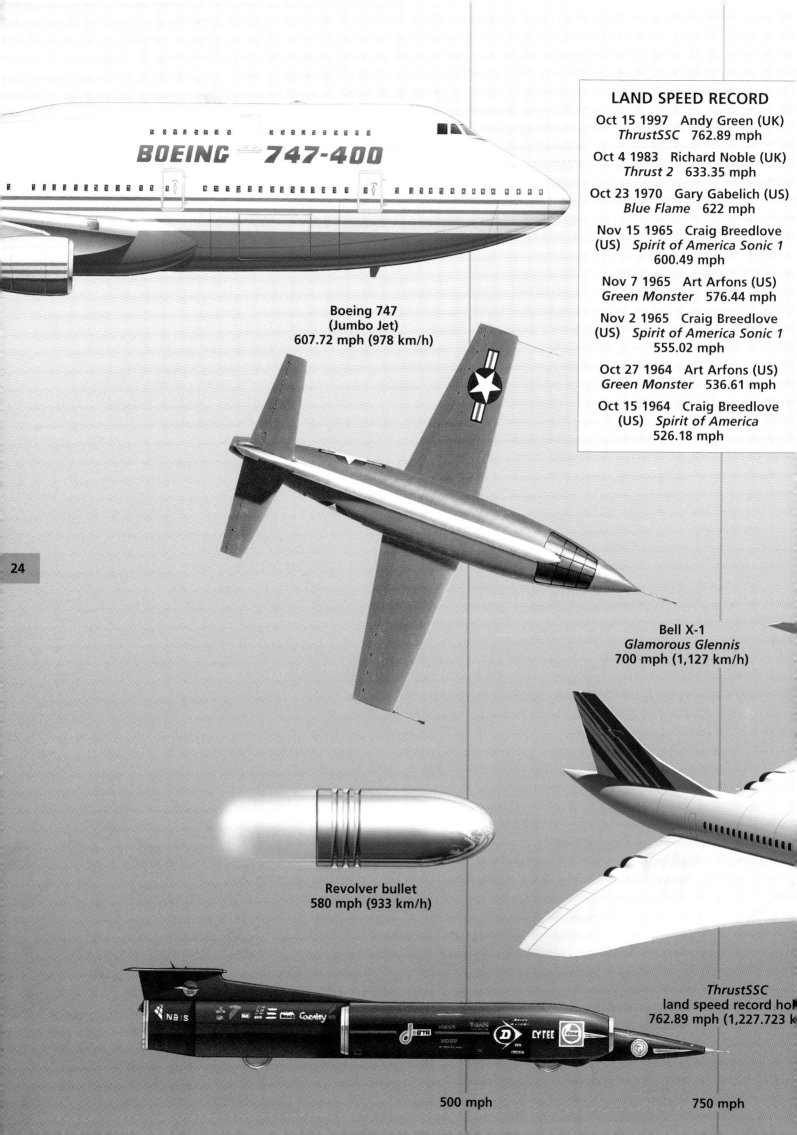

580 – 1,523 mph

ON THESE PAGES, we pass the speed of sound: about 761 mph (1,225 km/h) at sea level, or Mach 1. All forms of transport that go faster than this are called supersonic. The first aircraft to fly at supersonic speeds was the Bell X-1. It was piloted by U.S. pilot Chuck Yeager on October 14, 1947. Named *Glamorous Glennis* for Yeager's wife, it actually reached a speed of 700 mph (1,127 km/h), but the speed of sound is lower at high altitudes. Most modern fighter jet aircraft, such as the F-16, can fly much faster than Mach 1.

The first supersonic airliner, the Russian Tupolev-144, flew in 1968. But Concorde is the only supersonic airliner flying today. Cruising at 59,000 feet (about 18,000 m), it flies at more than twice the speed of sound (Mach 2). Other airliners, such as the Boeing 747, are known as "subsonic."

A land vehicle went faster than the speed of sound officially for the first time in 1997. *ThrustSSC*, driven by Andy Green, had two huge jet engines.

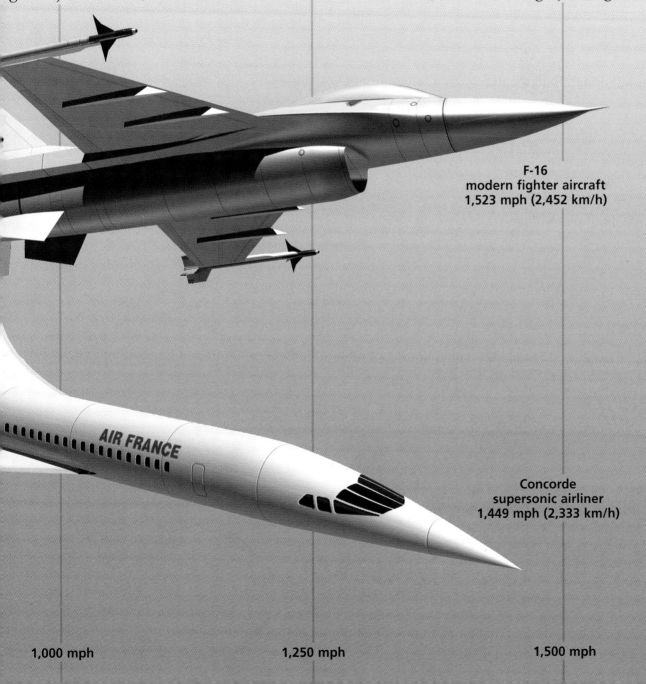

F-16
modern fighter aircraft
1,523 mph (2,452 km/h)

Concorde
supersonic airliner
1,449 mph (2,333 km/h)

1,000 mph   1,250 mph   1,500 mph

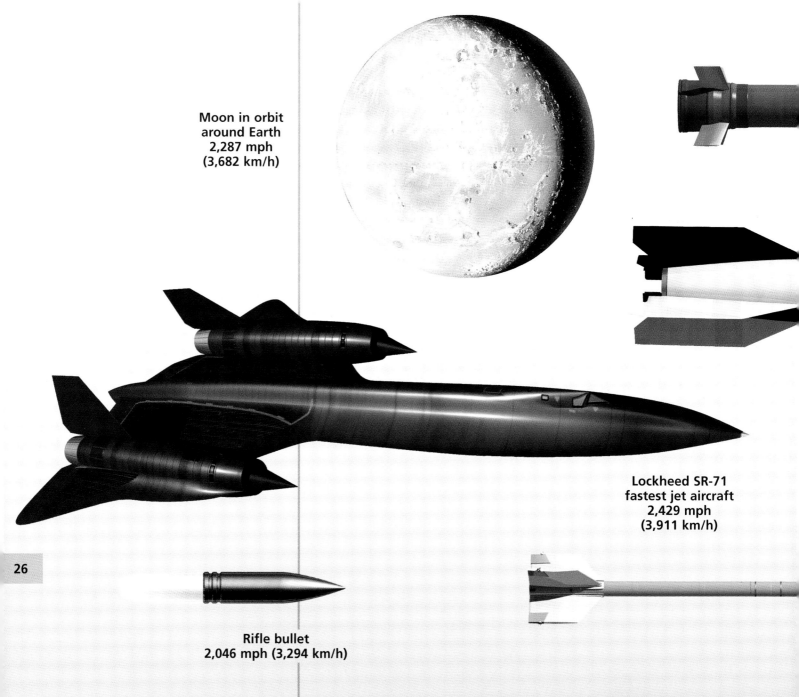

Moon in orbit around Earth
2,287 mph (3,682 km/h)

Lockheed SR-71 fastest jet aircraft
2,429 mph (3,911 km/h)

Rifle bullet
2,046 mph (3,294 km/h)

THE WORLD'S FASTEST AIRCRAFT

	plane	top speed	maximum altitude
1	X-15	Mach 6.72	354,109 feet
2	Lockheed SR-71	Mach 3.35	82,000 feet
3	X-2	Mach 3.12	126,000 feet
4	XB-70 *Valkyrie*	Mach 3.1	95,000 feet
5	MiG-31 *Foxhound*	Mach 2.83	69,000 feet
6	MiG-25 *Foxbat*	Mach 2.8	68,000 feet
7	F-15 *Eagle*	Mach 2.5	59,000 feet
8=	F-14A *Tomcat*	Mach 2.4	49,000 feet
8=	F-111 *Aardvark*	Mach 2.4	46,000 feet
10=	MiG-23 *Flogger*	Mach 2.35	59,000 feet
10=	Su-27 *Flanker*	Mach 2.35	56,000 feet
12	F-106 *Delta Dart*	Mach 2.31	52,500 feet

2,000 mph   2,500

2,046 – 4,518 mph

Anti-tank projectile
2,725 mph
(4,388 km/h)

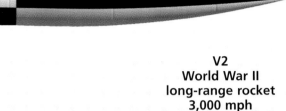

V2
World War II
long-range rocket
3,000 mph
(4,828 km/h)

Sidewinder
surface-to-air
missile
2,600 mph
(4,186 km/h)

WE ARE NOW looking at things that travel at extreme speed: aircraft, rockets, and missiles that can overtake a rifle bullet. These things can travel faster than the Moon in its orbit around the Earth.

Although it is more than 30 years old, the U.S. spyplane Lockheed SR-71 is still the world's fastest jet aircraft. Flying at an altitude of 82,000 feet (25,000 m), it can easily go faster than Mach 3 (three times the speed of sound). At such high speeds, its surface reaches 570°F (300°C)!

The fastest aircraft of all is the American experimental plane, the X-15. It is carried into the air attached to the underside of a transporter plane. Then its rocket engines blast it off. Not only is the X-15 the first aircraft to go faster than Mach 4, 5, and 6, it also holds the altitude record of 354,109 feet (107,960 m)—on the edge of space itself.

X-15
fastest
aircraft
4,518 mph
(7,274 km/h)

3,000 mph   4,000 mph   5,000 mph

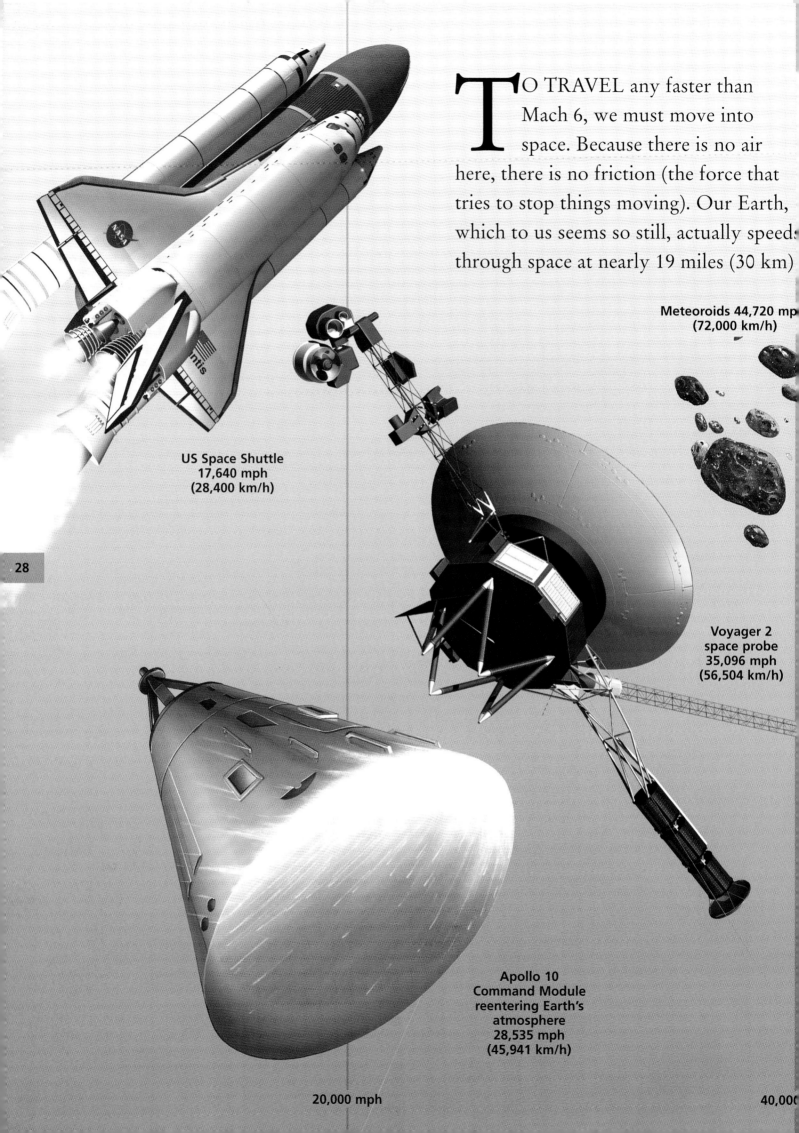

TO TRAVEL any faster than Mach 6, we must move into space. Because there is no air here, there is no friction (the force that tries to stop things moving). Our Earth, which to us seems so still, actually speeds through space at nearly 19 miles (30 km)

Meteoroids 44,720 mph (72,000 km/h)

US Space Shuttle 17,640 mph (28,400 km/h)

Voyager 2 space probe 35,096 mph (56,504 km/h)

Apollo 10 Command Module reentering Earth's atmosphere 28,535 mph (45,941 km/h)

20,000 mph   40,000

17,640 – 107,106 mph

er second. It is not the fastest planet, however. That record goes to Mercury, the nearest planet to the Sun. It orbits the Sun in just 88 days.

Meteoroids, small lumps of rock that fly around in space, go faster than the space probe Voyager 2. This unmanned spacecraft, was launched in 1977 on a journey to fly close by the giant planets, Jupiter, Saturn, Uranus, and Neptune, and to take detailed photographs of them. The fastest humanmade object of all time, Voyager 2 has since left the Solar System. It still sends back signals to Earth from many billions of miles away!

To travel into space, a vehicle must go at a speed of at least 17,640 mph (28,400 km/h). Only modern rockets, such as those used on the Space Shuttle, have enough power to do this.

Earth 66,600 mph (107,226 km/h)

Mercury 107,106 mph (172,440 km/h)

	HOW FAST ARE THE PLANETS?	
1	Mercury	30 miles per second
2	Venus	22 miles per second
3	Earth	19 miles per second
4	Mars	15 miles per second
5	Jupiter	8 miles per second
6	Saturn	6 miles per second
7	Uranus	4 miles per second
8	Neptune	3.5 miles per second
9	Pluto	2.8 miles per second

60,000 mph

INDEX

A
aircraft 14, 23, 24-25, 26-27
   light 21
   supersonic 25, 26-27
   world's fastest 26-27
airliner 25
   supersonic 25
airships 19
Albertosaurus 11
amoeba 6
animals, world's fastest land 16-17, 18-19
antelope, pronghorn 18
anti-tank projectile 27
Apollo 10 Command Module 28

B
Bell X-1 24-25
birds, world's fastest 18
Boeing 747 24-25
bullet,
  revolver 24
  rifle 26-27

C
car, Benz early petrol engine 9
caravels 10
cheetah 18-19
cockroach 7
Concorde 23, 25
crocodile 9
cyclist, racing 16

D
dinosaurs, fastest 11
dragonfly 12
dragster 23

E
Earth 28-29
Easyrider motorcycle 23
elephant, African 13
escape velocity 29

F
F-16 25
falcon, peregrine 18, 20
Formula 1 racing car 22
friction 28

G
gazelle, Thomson's 16-17, 18
Glamorous Glennis see Bell X-1
golf ball drive 22
gravity 29
greyhound 16

H
hare, brown 16-17
helicopters 23
Hindenburg 19, 23
honeybee 10

IJK
ice skater, speed 14
insects, fastest flying 12
Jai alai ball 22
kangaroo, red 16-17

L
land speed record 24
Lockheed SR-71 23, 26-27
longbow arrow 23
luge 19
Lynx helicopter 22-23

M
mallard duck 18
Mallard steam locomotive 20-21
Mercury 28-29
meteoroids 28-29
missiles 27
Moon 26-27
motorcycle, world's fastest 23

OP
ocean liner 13
ostrich 16-17, 18
pig 10
pigeon, racing 18
planets, speed of 29
polar bear 8, 15
powerboat 21

R
racehorse 16, 18
rat 9

reptiles, world's fastest 9
rhinoceros 15
Rocket 14
rockets 27, 29

S
Santa Maria 10
shark, tiger 15
Sidewinder surface-to-air missil 27
skateboarder 18
skier, downhill 21
sloth, three-toed 7
snail, garden 6
sound, speed of 25
Space Shuttle 28-29
spacecraft 29
Spirit of Australia 21, 23
sprinters 12-13
sprinting swimmer 8
stagecoach 13
steam locomotive 14, 20
   world's fastest 20-21
submarine, Alfa class 17
swift, spine-tailed 18, 20

T
table tennis smash 20
tennis serve 21
ThrustSSC 24-25
Titanic 13
tornado 18
tortoise, Galápagos giant 7
Train à Grande Vitesse (TGV) 22-23
trout, sea 11
tuna, bluefin 17
Tyrannosaurus rex 11

VW
V2 rocket 27
Voyager 2 space probe 28-29
water speed record 21, 23
windsurfer 16-17
woodcock, American 8
Wright *Flyer* 14

XY
X-15 26-27
yacht, racing 15